Metrication for the Modeller

Published by
Model & Allied Publications Ltd
13/35 Bridge Street
HEMEL HEMPSTEAD
Herts

Contents

Printed in Great Britain by
Page Bros (Norwich) Ltd., Norwich

1. Metrication—What It Means

In 1965 the Board of Trade came out in favour of adopting the metric system in the United Kingdom. The time scale visualised for the changeover was ten years—Britain being 'all metric' by 1975. The result has been some considerable confusion. Some industries 'went metric' at an early stage—notably the timber trade. Others have held back; and still others remain strongly opposed to metrication (especially those where American practice has a considerable influence). Many have compromised on a sort of 'bi-lingual' system, quoting both English and metric units in specifications.

For a considerable period yet, therefore, both English and metric units are likely to continue in use separately, outside of academic institutions. This applies particularly in the case of modelling and model engineering where the bulk of the published literature, plans and specifications are in English units. Also the model engineer's stock of tools are usually based on English standards, with additional metric sizes where necessary. No statutory ruling will cause these to become obsolete. They still remain as useful as before. It is also up to the individual—far more so than in industry proper, which is largely open to commercial and political influence—as to whether he continues to think and work in English units; takes the plunge and goes 'all metric' (converting his English standard tools, etc., in dimensional values); or aims to work with both systems, as appropriate.

The latter can become increasingly necessary as 'metrication' spreads and one finds oneself working more with metric plans and specifications. Here the simplest answer is usually to work in metric, for example use a metric rule to derive measurements in millimetres. If necessary, final dimensions can be converted into equivalent English sizes—e.g. to match the availability of materials.

This is simple conversion, and can work both ways. The main object of this book is to provide you with figures and information for making such conversions. At the same time it aims to go a little further and explain the relationship between the various alternative units which crop up in both systems, with the particular object of making *sense* of conversions. Simply look up the appropriate section—Linear Measure, Areas and Square Measure, or whatever you are involved with, and find all the information you should need in condensed form. It is hoped that this will be a lot quicker than, and just as useful, as 'official' guides and conversion tables. The British Standards Institution, for example, publish a 580 page book just on 'Metric Standards for Engineering', as well as a dozen other booklets on conversion factors and tables, recommendations and so on. Life is too short to wade through these, unless it becomes strictly necessary.

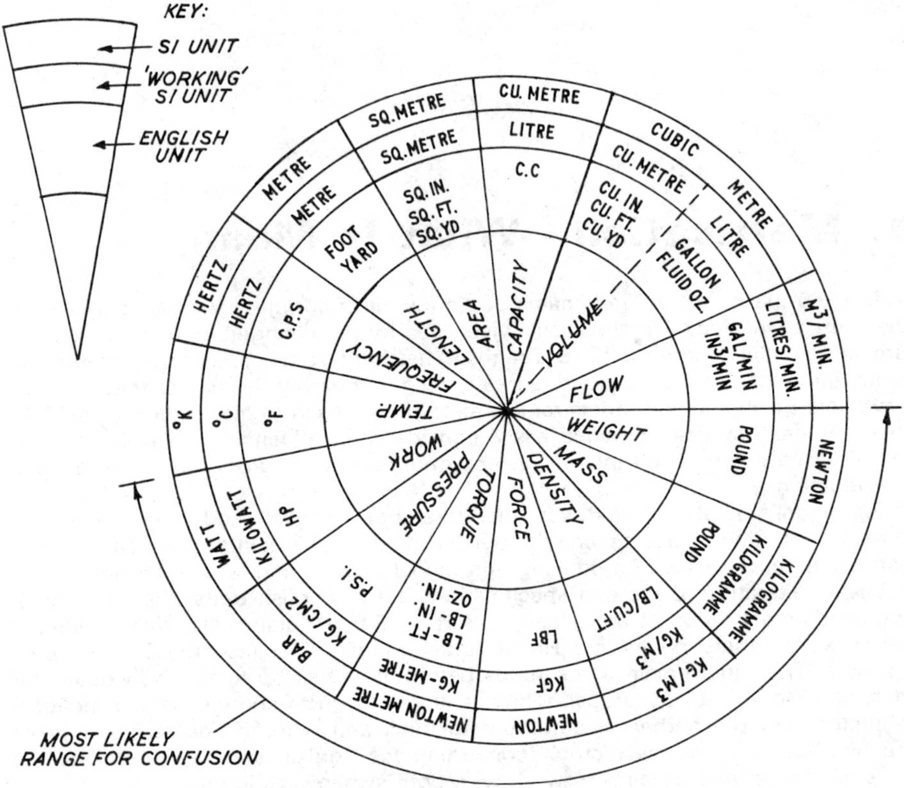

Further to assist in providing rapid and simple solutions, each section has its own set of 'instant conversion' scales, covering the more common units involved. These will be found most useful for general work as conversions can be read off the appropriate scales instantly, without calculation. Only a simple adjustment of a decimal point is necessary to accommodate values higher or lower than those covered by the scales.

Strangely enough much of the need for the profusion of explanations, new standards, etc, arises from the fact that the international system, known as the ISO system (after the International Organisation for Standardisation), is not the true original metric system as we know it at all. It is based on the SI system of units (Système Internationale d'Unités) which reduces the actual number of units required to six, covering all the technologies. The aim is to greatly simplify the relationship between units by eliminating arbitrary units, which are sometimes differently defined.

These six primary units are:

quantity	*unit*	*symbol*
LENGTH	metre	m
MASS	kilogramme	kg
TIME	seconds	s
ELECTRIC CURRENT	ampere	A
TEMPERATURE	degree Kelvin	°K
LUMINOUS INTENSITY	candela	cd

Most of these are already familiar. The supplementary and derived units, however, can lead to difficulties. In some cases they lead to 'unrealistic' large or small numerical values, when multiples and sub-multiples are preferred, in the same manner as the (technical) metric system—e.g. kilo for 1000 times, or milli for 1/1000th—See Appendix I. Also in some cases the new supplementary unit is given an unfamiliar name—e.g. **Newton** as the standard unit for **force.** But even the SI system 'bends' to the extent of accommodating 'customary' or 'workable' units; and in fact the majority of **working** units in the SI system are those already familiar in the technical metric system—see also the circular chart.

A Note About the Word 'Metrication'

Even the choice of word—'metrication' is open to some objection. Logically—in the engineering sense (which world is most affected by metrication)—the word should be 'metrification'. Whoever heard of 'electricating' a railway system, for example (or 'electrificating' it either, for that matter!)

Equally, from the point of view of dictionary English, a more correct word would be 'metricization' . . . or even 'metricize'. However, the word 'metrication' has been chosen by the Ministry of Technology on grounds of euphony—and 'metrication' it will remain!

2. Linear Measure

Conversions need rationalising since there is no exact equivalent between metric and English units, except for one:

1 yard = 0·9144 metre, exactly

The real trouble arises in considering inch fractions and their equivalents, as commonly used for standard balsa and other wood sizes. We really need two sets of *rationalised* equivalents, where the realistic or *actual* size is expressed either in English or metric units, and the corresponding equivalents in the other unit are *nominal* size. These can be based on 'logical' sizes, or those already in common use. The equivalent of $\frac{3}{4}$ in. for example, is 19·1 mm. Now no Continental worker would consider 19 mm as a 'logical' size. Instead he would adopt 20 mm as both a logical and practical equivalent.

This rounding off to logical figures can lead to some discrepancies. For example, $\frac{1}{4}$ in. works out as a 'difficult' size. The metric equivalent is 6·4 mm, which rounded off gives 6 mm. This is too near the rounded equivalent for $\frac{3}{16}$ in. = 5 mm. That makes $\frac{1}{4}$ in. an 'illogical' size in metric. The logical step up from 5 mm would be 7·5 mm, or approximately $\frac{5}{16}$ in. But $\frac{1}{4}$ in. is such an established dimension it has to remain available in material stock.

Largely on account of such discrepancies, and the multiplicity of published plans in English units, British manufacturers of sheet, strip and block wood for modelling are continuing to cut to actual inch sizes for the home market. Metric sizes are cut for the Continental market, but even here demand varies. Some countries accept 'English' lengths and 'metric' widths; others specify 'metric' lengths but take widths and thicknesses in 'English' units; whilst only a proportion are 'all metric'. This is quite different to the British timber trade which has gone 'all metric', but the problems of 'logical' sizing is not so apparent here because finished wood sizes in this case are only nominal.

The best thing to do with model sizes of wood is to work to one *rationalised* standard or the other, according to availability. Actual sizes used will then be exact, within normal manufacturing tolerances. The following Tables give these rationalised dimensions. For *exact* conversions of English to metric dimensions, and vice versa, consult Tables I, II and III.

Rationalised English Standard Sizes for Balsa etc.

Strip Sizes and Sheet Thicknesses

in. (actual)	mm (nominal)
$\frac{1}{32}$	0·8
($\frac{3}{64}$)	1
$\frac{1}{16}$	1·5
$\frac{3}{32}$	2·5
$\frac{1}{8}$	3
($\frac{5}{32}$)	(4)
$\frac{3}{16}$	5
$\frac{1}{4}$	6
($\frac{5}{16}$)	7·5
$\frac{3}{8}$	10
$\frac{1}{2}$	12·5
$\frac{5}{8}$	15
$\frac{3}{4}$	20
$\frac{7}{8}$	22·5
1	25
$1\frac{1}{4}$	30
$1\frac{1}{2}$	40
$1\frac{3}{4}$	45
2	50
$2\frac{1}{2}$	65
3	75

Widths

in. (actual)	mm (nominal)
1	25
2	50
3	75
4	100

Lengths

in. (actual)	mm (nominal)
12	300
24	610
36	915
48	1220

OTHER RATIONALISED DIMENSIONS

in. (actual)	mm (nominal)
5	125
6	150
7	175
8	200
9	225
10	250
11	275
15	375
18	450
20	500

( ) Indicate non-standard production sizes

Rationalised Metric Standard Sizes for Balsa etc.

Strip Sizes and Sheet Thicknesses

mm (actual)	in. (nominal)
0·8	$\frac{1}{32}$
1·5	$\frac{1}{16}$
2·5	$\frac{3}{32}$
3	$\frac{1}{8}$
5	$\frac{3}{16}$
(6)	($\frac{1}{4}$)
7·5	$\frac{5}{16}$
10	$\frac{3}{8}$
12·5	$\frac{1}{2}$
15	$\frac{5}{8}$
20	$\frac{3}{4}$
22·5	$\frac{7}{8}$
25	1
30	$1\frac{1}{9}$
35	$1\frac{3}{8}$
40	$1\frac{1}{2}$
50	2

Widths

mm (actual)	in. (nominal)
25	1
50	2
75	3
(80)	(3)
100	4

Lengths

mm (actual)	in. (nominal)
250	10
300	12
400	15
500	20
1000	40

OTHER RATIONALISED DIMENSIONS

mm (actual)	in. (nominal)
150	6
200	7·5
350	12·5
450	17·5
600	22·5
700	25
800	30
900	35

Wire and Sheet Gauges

The ISO recommendation is a series of millimetre sizes based on preferred numbers to take the place of all gauge numbers for specifying wire diameters and sheet metal thicknesses. Meantime, however, it seems almost certain that the established gauges (e.g. SWG in Britain and the Brown and Sharpe gauge in America) will continue to be used for a number of years—and likewise the original 'metric gauge', again based on gauge numbers.

ISO recommendations are based on three series, with preference given to sizes in the R10, R20 and R40 series, in that order. Corresponding dimensions are shown below, together with SWG equivalents or near-equivalents.

ISO Series size in mm R10	R20	R40	approximate swg equivalent
0·020	0·020	0·020	
		0·021	
	0·022	0·022	
		0·024	
0·025	0·025	0·025	50
		0·026	50
	0·028	0·028	
		0·030	49
0·032	0·032	0·032	49
		0·034	
	0·036	0·036	
		0·038	
0·040	0·040	0·040	48
		0·042	
	0·045	0·045	
		0·048	
0·050	0·050	0·050	47
		0·053	
	0·056	0·056	
		0·060	46
0·063	0·063	0·063	
		0·067	
	0·071	0·071	45*
		0·075	
0·080	0·080	0·080	44
		0·085	
	0·090	0·090	43
		0·095	
0·100	0·100	0·100	42
		0·106	
	0·112	0·112	41
		0·118	
0·125	0·125	0·125	
		0·132	39*
	0·140	0·140	
		0·150	38
0·160	0·160	0·160	
		0·170	37
	0·180	0·180	
		0·190	36
0·200	0·200	0·200	
		0·212	35
	0·224	0·224	
		0·236	34
0·250	0·250	0·250	33
		0·265	
	0·280	0·280	32
		0·300	31
0·315	0·315	0·315	30*
		0·335	
	0·355	0·355	
		0·375	28*
0·400	0·400	0·400	
		0·425	
	0·450	0·450	26
		0·475	
0·500	0·500	0·500	
		0·530	
	0·560	0·560	24*
		0·600	
0·630	0·630	0·630	
		0·670	
	0·710	0·710	22*
		0·750	
0·800	0·800	0·800	21

ISO Series size in mm R10	R20	R40	approximate swg equivalent
		0·850	
	0·900	0·900	20
		0·950	
1·00	1·00	1·00	19
		1·06	
	1·12	1·12	
		1·18	18
1·25	1·25	1·25	
		1·32	
	1·40	1·40	17
		1·50	
1·60	1·60	1·60	16
		1·70	
	1·80	1·80	15
		1·90	
2·00	2·00	2·00	14
		2·12	
	2·24	2·24	
		2·36	13
2·50	2·50	2·50	
		2·65	12*
	2·80	2·80	
		3·00	11

ISO Series size in mm RZ0	R20	R40	approximate swg equivalent
3·15	3·15	3·15 }	10
		3·35 }	
	3·55	3·55 }	9
		3·75 }	
4·00	4·00	4·00	8
		4·25	
	4·50	4·50	7
		4·75	6
5·00	5·00	5·00	
		5·30	5
	5·60	5·60	
		6·00	4
6·30	6·30	6·30	3
		6·70	
	7·10	7·10	2
		7·50	1
8·00	8·00	8·00	0
		8·50	
	9·00	9·00	
		9·50	
10·00	10·00	10·00	

* More or less exact

Drill Sizes

Number and Letter drill sizes have been obsolete for a number of years. That does not make stocks of such drills any the less useful. What really matters is knowing the equivalents in other standard sizes which may be specified on new designs, i.e. millimetre (metric), or inch size in decimals. The following Table is a complete guide.

Inches decimals	in.	MM	Old No. or Letter sizes	Inches decimals	in.	MM	Old No. or Letter sizes
0·0126		0·32		0·0492		1·25	
0·0138		0·35	80	0·0512		1·30	55
0·0150		0·38	79	0·0532		1·35	
0·0156	$\frac{1}{64}$			0·0551		1·40	54
0·0157		0·40	78	0·0571		1·45	

Inches decimals	in.	MM	Old No. or Letter sizes	Inches decimals	in.	MM	Old No. or Letter sizes
0·0165		0·42		0·0590		1·50	53
0·0177		0·45	77	0·0610		1·55	
0·0189		0·48		0·0625	$\frac{1}{16}$		
0·0197		0·50	76	0·0630		1·60	52
0·0205		0·52	75	0·0650		1·65	
0·0216		0·55		0·0669		1·70	51
0·0228		0·58	74	0·0689		1·75	
0·0236		0·60	73	0·0709		1·80	50
0·0244		0·62		0·0728		1·85	49
0·0256		0·65	72 & 71	0·0748		1·90	
0·0268		0·68		0·0768		1·95	48
0·0276		0·70	70	0·0781	$\frac{5}{64}$		
0·0283		0·72		0·0787		2·00	47
0·0295		0·75	69	0·0807		2·05	46
0·0307		0·78		0·0827		2·10	45
0·0312	$\frac{1}{32}$		68	0·0846		2·15	
0·0315		0·80		0·0866		2·20	44
0·0323		0·82	67	0·0886		2·25	43
0·0335		0·85	66	0·0906		2·30	
0·0346		0·88		0·0925		2·35	
0·0354		0·90	65	0·0938	$\frac{3}{32}$		42
0·0362		0·92	64	0·0945		2·40	
0·0374		0·95	63	0·0965		2·45	41
0·0386		0·98	62	0·0984		2·50	40
0·0394		1·00	61 & 60	0·1004		2·55	39
0·0413		1·05	59 & 58	0·1024		2·60	38
0·0433		1·10	57	0·1043		2·65	37
0·0453		1·15		0·1063		2·70	36
0·0469	$\frac{3}{64}$		56	0·1083		2·75	
0·0472		1·20		0·0194	$\frac{7}{64}$		
0·1102		2·80	35 & 34	0·2165		5·50	
0·1122		2·85	33	0·2188	$\frac{7}{32}$		
0·1142		2·90		0·2205		5·60	2
0·1161		2·95	32	0·2244		5·70	
0·1181		3·00	31	0·2283		5·80	1
0·1220		3·10		0·2323		5·90	
0·1250	$\frac{1}{8}$			0·2344	$\frac{15}{64}$		A
0·1260		3·20		0·2362		6·00	B
0·1299		3·30	30	0·2402		6·10	C
0·1339		3·40		0·2441		6·20	D

Inches decimals	in.	MM	Old No. or Letter sizes	Inches decimals	in.	MM	Old No. orLetter sizes
0·1378		3·50	29	0·2480		6·30	
0·1406	$\frac{9}{64}$		28	0·2500	$\frac{1}{4}$		E
0·1417		3·60		0·2520		6·40	
0·1457		3·70	27 & 26	0·2559		6·50	F
0·1496		3·80	25	0·2598		6·60	G
0·1535		3·90	24 & 23	0·2638		6·70	
0·1562	$\frac{5}{32}$			0·2656	$\frac{17}{64}$		H
0·1575		4·00	22 & 21	0·2677		6·80	
0·1614		4·10	20	0·2717		6·90	I
0·1654		4·20	19	0·2756		7·00	J
0·1693		4·30	18	0·2795		7·10	
0·1719	$\frac{11}{64}$			0·2812	$\frac{9}{32}$		K
0·1732		4·40	17	0·2835		7·20	
0·1772		4·50	16	0·2874		7·30	
0·1811		4·60	15 & 14	0·2913		7·40	L
0·1850		4·70	13	0·2953		7·50	M
0·1875	$\frac{3}{16}$			0·2969	$\frac{19}{64}$		
0·1890		4·80	12	0·2992		7·60	
0·1929		4·90	11 & 10	0·3031		7·70	N
0·1968		5·00	9	0·3071		7·80	
0·2008		5·10	8 & 7	0·3110		7·90	
0·2031	$\frac{13}{64}$			0·3125	$\frac{5}{16}$		
0·2047		5·20	6 & 5	0·3150		8·00	O
0·2087		5·30	4	0·3189		8·10	
0·2126		5·40	3	0·3228		8·20	P

TABLE I. MILLIMETRES TO INCHES

mm	0	1	2	3	4	5	6	7	8	9
0	—	0·0394	0·0787	0·1181	0·1575	0·1969	0·2362	0·2756	0·3150	0·3543
10	0·3937	0·4331	0·4724	0·5118	0·5512	0·5906	0·6299	0·6693	0·7087	0·7480
20	0·7874	0·8268	0·8661	0·9055	0·9449	0·9843	1·0236	1·0630	1·1024	1·1417
30	1·1811	1·2205	1·2598	1·2992	1·3386	1·3780	1·4173	1·4570	1·4961	1·5354
40	1·5748	1·6142	1·6535	1·6929	1·7323	1·7717	1·8110	1·8504	1·8898	1·9291
50	1·9685	2·0079	2·0472	2·0866	2·1260	2·1654	2·2047	2·2441	2·2835	2·3228
60	2·3622	2·4016	2·4409	2·4803	2·5197	2·5591	2·5984	2·6378	2·6772	2·7165
70	2·7559	2·7953	2·8347	2·8740	2·9134	2·9528	2·9921	3·0315	3·0709	3·1102
80	3·1496	3·1890	3·2284	3·2677	3·3071	3·3465	3·3858	3·4252	3·4646	3·5039
90	3·5433	3·5827	3·6221	3·6614	3·7008	3·7402	3·7795	3·8189	3·8583	3·8976

TABLE II. INCHES AND INCH FRACTIONS TO MILLIMETRES

Inch fraction	Inches											
	0	1	2	3	4	5	6	7	8	9	10	11
—	—	25·4	50·8	76·2	101·6	127·0	152·4	177·8	203·2	228·6	254·0	279·
$\frac{1}{16}$	1·6	27·0	52·4	77·8	103·2	128·6	154·0	179·4	204·8	230·2	255·6	281·
$\frac{1}{8}$	3·2	28·6	54·0	79·4	104·8	130·2	155·6	181·0	206·0	231·8	257·2	282·
$\frac{3}{16}$	4·8	30·2	55·6	81·0	106·4	131·8	157·2	182·6	208·0	233·4	258·8	284·
$\frac{1}{4}$	6·4	31·8	57·2	82·6	108·0	133·4	158·8	184·2	209·6	235·0	260·4	285·
$\frac{5}{16}$	7·9	33·3	58·7	84·1	109·5	134·9	160·3	185·7	211·1	236·5	261·9	287·
$\frac{3}{8}$	9·5	34·9	60·3	85·7	111·1	136·5	161·9	187·3	212·7	238·1	263·5	288·
$\frac{7}{16}$	11·1	36·5	61·9	87·3	112·7	138·1	163·5	188·9	214·3	239·7	265·1	290·
$\frac{1}{2}$	12·7	38·1	63·5	88·9	114·3	138·7	165·1	190·5	215·9	241·3	266·7	292·
$\frac{9}{16}$	14·3	39·7	65·1	90·5	115·9	141·3	166·7	192·1	217·5	242·9	268·3	293·
$\frac{5}{8}$	15·9	41·3	66·7	92·1	117·5	142·9	168·3	193·7	219·1	244·5	269·9	295·
$\frac{11}{16}$	17·5	42·9	68·3	93·7	119·1	144·5	169·9	195·3	220·7	246·1	271·5	296
$\frac{3}{4}$	19·1	44·5	69·9	95·3	120·7	146·1	171·5	196·9	222·3	247·7	273·1	298·
$\frac{13}{16}$	20·6	46·0	71·4	96·8	122·2	147·6	173·0	198·4	223·8	249·2	274·6	300·
$\frac{7}{8}$	22·2	47·6	73·0	98·4	123·8	149·2	174·6	200·0	225·4	250·8	276·2	301·
$\frac{15}{16}$	23·8	49·2	74·6	100·0	125·4	150·8	176·2	201·6	227·0	252·4	277·8	303·

TABLE IIIA. FEET AND INCHES TO METRES

Inches	Feet										
	0	1	2	3	4	5	6	7	8	9	10
—	—	0·3048	0·6096	0·9144	1·2192	1·5240	1·8288	2·1336	2·4384	2·7432	3·0480
1	0·0254	0·3302	0·6350	0·9398	1·2446	1·5494	1·8542	2·1590	2·4638	2·7686	3·0734
2	0·0508	0·3556	0·6604	0·9652	1·2700	1·5748	1·8796	2·1844	2·4892	2·7940	3·0988
3	0·0762	0·3810	0·6858	0·9906	1·2954	1·6002	1·9050	2·2098	2·5146	2·8194	3·1242
4	0·1016	0·4064	0·7112	1·0160	1·3208	1·6256	1·9304	2·2352	2·5400	2·8448	3·1496
5	0·1270	0·4318	0·7366	1·0414	1·3462	1·6510	1·9558	2·2606	2·2659	2·8702	3·1750
6	0·1524	0·4572	0·7620	1·0668	1·3716	1·6764	1·9812	2·2860	2·5908	2·8956	3·2004
7	0·1778	0·4826	0·7874	1·0922	1·3970	1·7018	2·0066	2·3114	2·6162	2·9210	3·2258
8	0·2032	0·5080	0·8128	1·1176	1·4224	1·7272	2·0320	2·3368	2·6416	2·9464	3·2512
9	0·2286	0·5334	0·8382	1·1430	1·4478	1·7526	2·0574	2·3622	2·6670	2·9718	3·2766
10	0·2540	0·5588	0·8636	1·1684	1·4732	1·7780	2·0828	2·3876	2·6924	2·9972	3·3020
11	0·2794	0·5842	0·8890	1·1938	1·4986	1·8034	2·1082	2·4130	2·7178	3·0226	3·3274

TABLE IIIB. METERS TO FEET

Metres	1	2	3	4	5	6	7	8	8	10
Feet	3·28	6·56	9·84	13·12	16·40	19·685	22·97	26·24	29·53	32·8

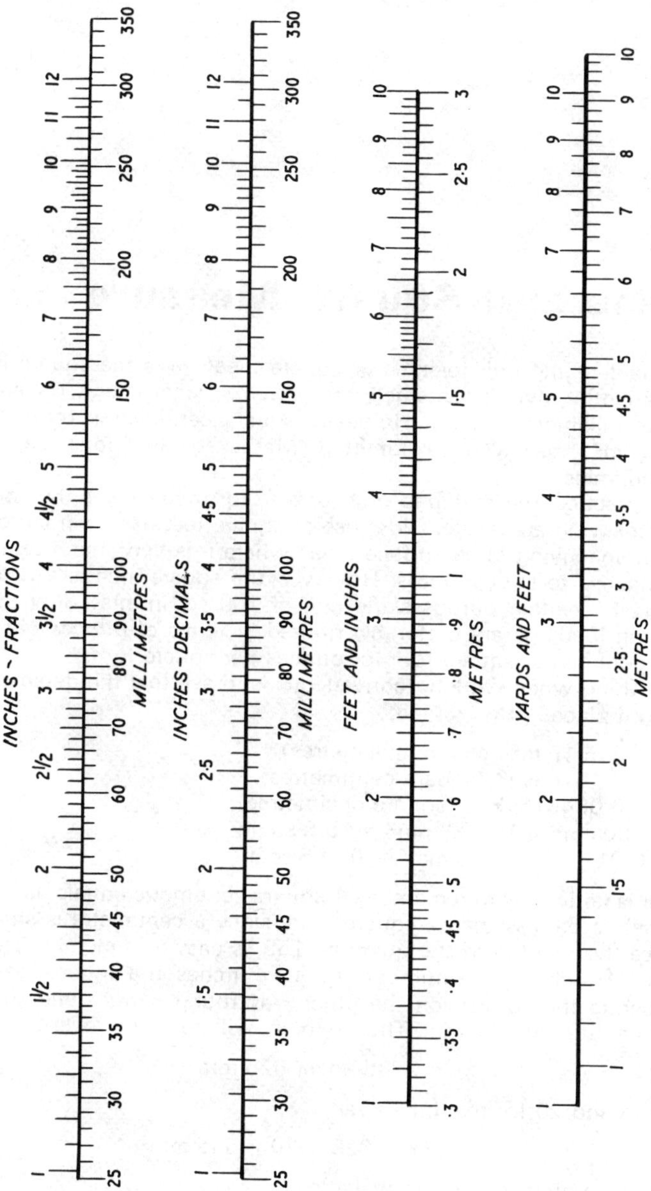
INCHES - FRACTIONS
MILLIMETRES
INCHES - DECIMALS
MILLIMETRES
FEET AND INCHES
METRES
YARDS AND FEET
METRES

3. Areas and Square Measure

The standard metric unit for area or square measure is the square metre (m^2). Where the quantity involved is small, the *preferred* smaller unit (or sub-multiple) is the square millimetre (mm^2). However, square centimetres (cm^2) and square decimetres (dm^2) may also be admitted. Metric specifications may use any of these sub-multiples.

In British practice, the preferred unit for area for modellers is the square inch—e.g. quoted for wing areas, etc. Older use of square feet has been largely dropped, except when specifying large surface areas where it is very much preferred in the engineering sense to square yards. However, the square inch is a relatively small quantity, and for certain purposes the unit of 100 sq. in. may often be used. For example, wing loadings are preferably quoted in terms of *ounces per 100 sq. in.*, rather than pounds per square foot, or ounces per square foot.

The main thing when working conversions is to see that the decimal point ends up in the right place.

1 sq. in. = 645·16 m^2 (square millimetres)
= 6·4516 cm^2 (square centimetres)
= 0·064516 dm^2 (square decimetres)
1 dm^2 = 100 cm^2 = 10 000 mm^2 = 15·5 sq. in.
1 cm^2 = 0·01 dm^2 = 100 mm^2 = 0·155 sq.in.

The latter is fairly easy to remember. 1 square centimetre equals *point* 155 square inches. It has to be less than a square inch, since a centimetre is smaller than an inch—hence the *point*. And the quantity '155' is easy to remember (just as one is familiar with '144' being the number of square inches in a square foot).

Remembering the conversion the other way round is not quite so easy. However, 1 in. is a little over 25 mm. Thus 1 sq. in. will be a little over 25 squared

= a little over 625 mm^2

Remember to add 20 for the 'little over'

1 sq. in. = 625 + 20 = 645 mm^2

That will give an almost exact conversion.

Alternatively, simply use Tables IV and V for looking up exact conversions.

TABLE IV A. SQUARE INCHES TO SQUARE CENTIMETRES

↓ sq. in.→	0	1	2	3	4	5	6	7	8	9
0	—	6·45	12·90	19·35	25·81	32·86	38·71	45·16	51·61	58·06
10	64·52	70·97	77·42	83·57	90·32	96·77	103·23	109·68	115·16	122·58
20	129·03	135·48	141·94	148·39	154·84	161·29	167·74	174·19	180·65	187·10
30	193·55	200·00	206·45	212·90	219·35	225·81	232·26	238·71	245·16	251·61
40	258·06	264·52	270·97	277·42	283·87	290·32	296·77	303·23	309·68	316·13
50	322·58	329·03	335·48	341·94	348·39	354·84	361·29	367·74	374·19	380·64
60	387·10	393·55	400·00	406·45	412·90	419·35	425·81	432·26	438·71	445·16
70	451·61	458·06	464·52	470·97	477·42	483·87	490·32	496·77	503·22	509·68
80	516·13	522·58	529·03	535·48	541·93	548·39	554·84	561·29	567·74	574·19
90	580·64	587·10	593·55	600·00	606·45	612·90	619·35	625·81	632·26	638·71
100	645·16	—	—	—	—	—	—	—	—	—

TABLE IV B. SQUARE CENTIMETRES TO SQUARE INCHES

↓ sq. cm→	0	1	2	3	4	5	6	7	8	9
0	—	0·155	0·310	0·465	0·620	0·775	0·930	1·085	1·240	1·395
10	1·550	1·705	1·860	2·015	2·170	2·325	2·480	2·635	2·790	2·945
20	3·100	3·255	3·410	3·568	3·720	3·875	4·030	4·185	4·340	4·495
30	4·650	4·805	4·960	5·115	5·270	5·425	5·580	5·735	5·890	6·045
40	6·200	6·355	6·510	6·665	6·820	6·975	7·130	7·285	7·440	7·595
50	7·750	7·905	8·060	8·215	8·370	8·525	8·680	8·835	8·990	9·145
60	9·300	9·455	9·610	9·765	9·920	10·075	10·230	10·385	10·540	10·695
70	10·850	11·005	11·160	11·315	11·470	11·625	11·780	11·935	12·090	12·245
80	12·400	12·595	12·710	12·865	13·020	13·175	13·330	13·485	13·640	13·795
90	13·950	14·105	14·260	14·415	14·570	14·725	14·880	15·035	15·190	15·345
100	15·500	—	—	—	—	—	—	—	—	—

TABLE V A. SQUARE FEET TO SQUARE METRES

↓ sq. ft. →	0	1	2	3	4	5	6	7	8	9
0	—	0·0929	0·186	0·279	0·372	0·465	0·557	0·650	0·793	0·836
10	0·929	1·022	1·115	1·208	1·301	1·394	1·486	1·579	1·672	1·765
20	1·858	1·951	2·044	2·137	2·230	2·323	2·415	2·508	2·601	2·694
30	2·787	2·880	2·973	3·066	3·159	3·252	3·345	3·437	3·530	3·623
40	3·716	3·809	3·902	3·995	4·088	4·181	4·274	4·366	4·459	4·552
50	4·645	4·738	4·831	4·924	5·017	5·110	5·203	5·296	5·388	5·481
60	5·574	5·667	5·760	5·853	5·946	6·039	6·132	6·225	6·317	6·410
70	6·503	6·596	6·689	6·782	6·875	6·968	7·061	7·154	7·246	7·339
80	7·432	7·525	7·618	7·711	7·804	7·897	7·990	8·083	8·176	8·268
90	8·361	8·454	8·547	8·640	8·733	8·826	8·919	9·012	9·105	9·197
100	9·290	—	—	—	—	—	—	—	—	—

TABLE V B. SQUARE METRES TO SQUARE FEET

↓ sq. m. →	0	1	2	3	4	5	6	7	8	9
0	—	10·76	21·53	32·29	43·06	53·82	64·58	75·35	86·11	96·88
20	107·6	118·4	129·2	139·9	150·7	161·5	172·2	183·0	193·8	204·5
20	215·3	226·0	236·8	247·6	258·3	269·1	279·9	290·6	301·4	312·2
30	322·9	333·7	344·4	355·2	366·0	376·7	387·5	398·3	409·0	419·8
40	430·6	441·3	452·1	462·8	473·6	484·4	495·1	509·9	516·7	527·4
80	538·2	549·0	559·7	570·5	581·3	592·0	602·8	613·5	624·3	635·1
60	645·8	656·6	667·4	678·1	688·9	699·7	710·4	721·2	732·0	742·7
70	753·5	764·2	775·0	785·8	796·5	807·3	818·1	828·8	839·6	850·4
80	861·1	871·9	882·6	893·4	904·2	914·9	925·7	936·5	947·2	958·0
90	968·8	979·5	990·3	1001·0	1011·8	1022·6	1033·3	1044·1	1054·9	1065·6
100	1076·4	—	—	—	—	—	—	—	—	—

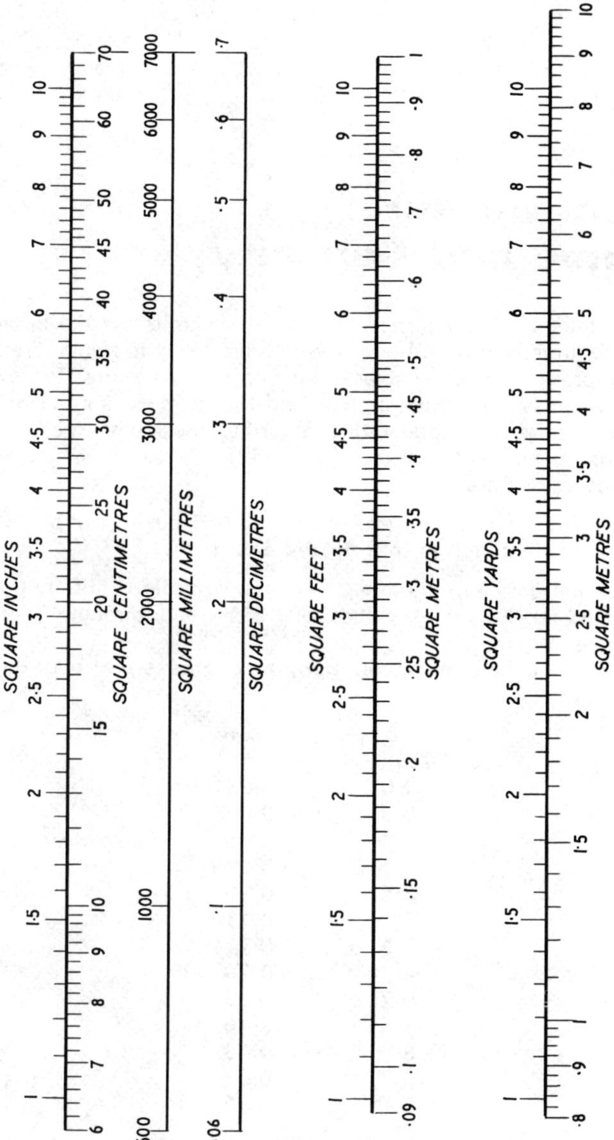
SQUARE INCHES
SQUARE CENTIMETRES
SQUARE MILLIMETRES
SQUARE DECIMETRES
SQUARE FEET
SQUARE METRES
SQUARE YARDS
SQUARE METRES

4. Cubic Measure: Volume and Capacity

As far as modellers are concerned the two main units are the cubic inch (in^3) and the cubic centimetre (cm^3), both of which are likely to remain in use for many years. The swept volume or displacement of glow motors, for example, will continue to be expressed in cubic inches; and that of diesels in c.c.'s. (cm^3). This has been current since the appearance of both types, with their American and European origins, respectively.

Standard conversions are:

1 cubic centimetre (1 cm^3) = 0·061 cu. in. (in^3)
1 cubic inch (1 in^3) = 16·4 c.c. (cm^3)

Both of these are easy to remember, but to save tedious arithmetic, exact conversions of mixed or fractional values are best read directly from conversion tables—see Tables VI and VII.

General equivalents are based on popular engine sizes, and these are also useful to remember.

c.c. (*cm^3*)	*cu. in.* (*in^3*)
0·5	0·031
0·8	**0·049**
1·0	0·061
1·6	**0·099**
2·5	**0·15**
3·1	**0·119**
3·5	0·214
4·75	**0·29**
5·75	**0·35**
7·5	**0·45**
8	**0·49**
10	**0·61**
15	0·92

Figures in **bold** are the particular 'standard' in each pair.

The ISO volumetric standard unit is the cubic metre (m^3)

TABLE VI. CUBIC INCHES TO CUBIC CENTIMETRES

cu. in.	0·0	0·01	0·02	0·03	0·04	0·05	0·06	0·07	0·08	0·09
—	—	0·16	0·33	0·49	0·66	0·82	0·98	1·15	1·31	1·47
0·1	1·64	1·80	1·97	2·13	2·29	2·46	2·62	2·79	2·95	3·11
0·2	3·28	3·44	3·61	3·77	3·93	4·10	4·26	4·42	4·59	4·75
0·3	4·29	5·08	5·24	5·41	5·57	5·74	5·90	6·06	6·23	6·39
0·4	6·55	6·72	6·88	7·05	7·21	7·37	7·54	7·70	7·87	8·03
0·5	8·19	8·36	8·52	8·69	8·85	9·01	9·18	9·34	9·50	9·67
0·6	9·83	10·00	10·16	10·32	10·49	10·65	10·82	10·98	11·14	11·31
0·7	11·47	11·63	11·80	11·96	12·13	12·29	12·45	12·62	12·78	12·95
0·8	13·11	13·27	13·44	13·60	13·77	13·93	14·09	14·26	14·42	14·58
0·9	14·75	14·91	15·08	15·24	15·40	15·57	15·73	15·90	16·06	16·22
1·0	16·39	—	—	—	—	—	—	—	—	—

TABLE VII. CUBIC CENTIMETRES TO CUBIC INCHES

c.c.	0·0	0·1	0·2	0·3	0·4	0·5	0·6	0·7	0·8	0·9
—	—	0·006	0·012	0·018	0·024	0·031	0·037	0·043	0·049	0·055
1	0·061	0·067	0·073	0·079	0·085	0·092	0·098	0·104	0·110	0·116
2	0·122	0·128	0·134	0·140	0·146	0·153	0·159	0·165	0·171	0·177
3	0·183	0·189	0·195	0·201	0·207	0·214	0·220	0·226	0·232	0·238
4	0·244	0·250	0·256	0·262	0·269	0·275	0·281	0·287	0·293	0·299
5	0·305	0·311	0·317	0·323	0·330	0·336	0·342	0·348	0·354	0·360
6	0·366	0·372	0·378	0·384	0·391	0·397	0·403	0·409	0·415	0·421
7	0·427	0·433	0·439	0·445	0·452	0·458	0·464	0·470	0·476	0·482
8	0·488	0·494	0·500	0·507	0·513	0·519	0·525	0·531	0·537	0·543
9	0·549	0·555	0·561	0·568	0·574	0·580	0·586	0·592	0·598	0·604
10	0·610	—	—	—	—	—	—	—	—	—

1 cubic metre = 1·308 cubic yards
= 35·315 cubic feet
= 61 000 cubic inches (approx)
1 cubic foot = 0·02832 cubic metres
1 cubic yard = 0·7646 cubic metres

These are conversions you may need from time to time, although the cubic metre is a very large unit and sub-multiples are more likely to be quoted. In case you find sub-multiples a little confusing:

1 cubic metre = 1000 cubic decimetres
1 cubic decimetre = 1000 cubic centimetres

Liquid Measure

The use of cubic metres as units for liquid volume is rather ridiculous for general applications, as are its sub-multiples. That brings us back to the *litre* as the rational metric unit (which, incidentally, is nominally a cubic *decimetre*; or 1000·028 cubic centimetres exactly); and alongside it the gallon, for as long as it survives.

1 litre = 0·219975 or near enough 0·22 gallons
1 gallon = 4·54596 litres = 0·16 cubic feet

This is the British or Imperial gallon. The American gallon is smaller.

1 US gallon = 0·83267 Imperial gallon = 3·8 litres approx.
1 Imperial gallon = 1·2 US gallons

As a rough approximation, take four-fifths of a US gallon to convert to Imperial gallons and then add 5 per cent (one twentieth). To convert Imperial gallons to US gallons, multiply by 6/5. That is exact to within one part in a thousand. This also applies when comparing US and British pints (both 8 to the gallon).

For smaller liquid measure the cubic centimetre is already more or less universally accepted, but under the IOS standard should be written cm^3, not the more familiar c.c.

The conversion to cubic inches (in^3) is given above, but the British have never been entirely fond of this English unit for small volumes. The *fluid ounce* is much preferred, which is 1/20th of a pint—quite an irrational unit really. The following should sort out the three 'small volume' units:

1 cubic centimetre = 0·035195 fluid ounces (say 0·0352)
= 0·061 cubic inches
1 cubic inch = 0·57674 fluid ounces = 16·4 cubic centimetres
1 fluid ounce = 28·413 (say 28·4) cubic centimetres
= 1·7339 (say 1·734) cubic inches

The fluid ounce, therefore, is approximately $1\frac{3}{4}$ cubic inches, or a little less than 30 cubic centimetres.

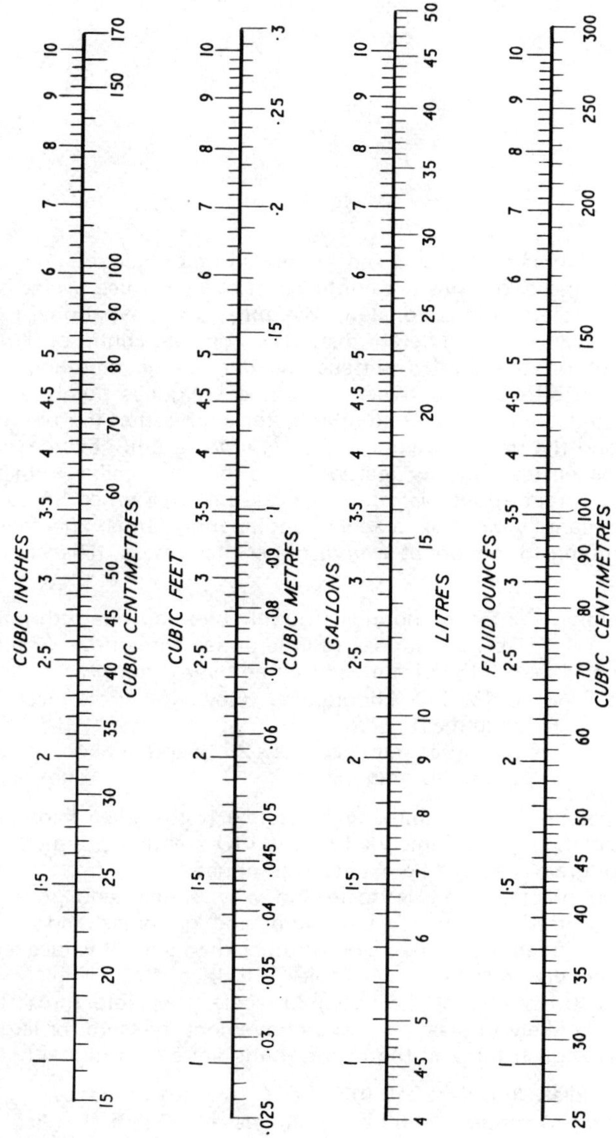
CUBIC INCHES
CUBIC CENTIMETRES
CUBIC FEET
CUBIC METRES
GALLONS
LITRES
FLUID OUNCES
CUBIC CENTIMETRES

5. Weights

The scientist differentiates between the *mass* of an object and its *weight* . . . and complain when the two terms are confused in everyday use. *Mass* is the quantity of matter in an object, and is constant. *Weight* is the gravitational force acting on the object, and varies with different distances from the centre of the Earth. At any particular point on the Earth's surface, *weight* is constant and so for normal practical use the difference between mass and weight is purely academic. What can be confusing under the ISO standards, though, is that the unit of mass is the *kilogramme*, and the unit of weight is the *Newton* (a unit of *force*).

Probably the easiest way to deal with this, unless familiar with the academic approach, is to forget about 'Newtons' and assume that when 'mass' is referred to it effectively means 'weight' in practical engineering terms; and there is really no harm in continuing to talk about *weight* rather than 'mass'. Then at least the units will be familiar.

1 kilogramme	= 2·20462 pounds (roughly two and one fifth pounds)
	= 35·274 ounces (a little under 'three dozen' ounces)
1 gramme	= 0·035274 ounces (a little over $\frac{1}{32}$nd of an ounce)
1 pound	= 0·453592 kilogramme (say 454 grammes—roughly 450 grammes)
1 ounce	= 28·3495 grammes (say 28·35 grammes—or roughly a little under 30 grammes)

Note the spelling of kilogramme and gramme (rather than kilogram and gram). The 'me' is specifically recommended by the ISO standards, although the spelling 'kilogram' and 'gram' can still be expected to appear.

In modelling practice there is no reason why ounces and pounds should not continue to be used for many years; alongside grammes and kilogrammes for Continental specifications. Conversion of other than simple values is best done by reference to conversion tables—see Tables VIII, IX and X.

Right at the other end of the 'weights' scale, there are three different 'tons' which are hardly likely to disappear as a convenient measure for large weights (or masses!). The English ton and the metric tonne are even rational near-equivalents

1 (English) ton = 2240 lb = 1·016 (metric) tonnes
1 (metric) tonne = 1000 kilogrammes = 0·99286 (English) ton

The American or 'short ton' is equal to 2000 pounds—roughly nine-tenths of the (English) ton or (metric) tonne.

TABLE VIII. OUNCES TO GRAMMES

(to 1 decimal place)

↓ ounces →	0·0	0·1	0·2	0·2	0·3	0·5	0·6	0·7	0·8	0·9
0	—	2·84	5·67	8·50	11·3	14·2	17·0	19·8	22·7	25·5
1	28·3	31·2	34·0	36·9	39·7	42·5	45·4	48·2	51·0	53·9
2	56·7	59·5	62·4	65·2	68·0	70·9	73·7	76·5	79·9	82·2
3	85·0	87·9	90·7	93·6	96·3	99·2	102·1	104·9	107·7	110·6
4	113·4	116·2	119·1	121·9	124·7	127·6	130·4	133·2	136·1	138·9
5	141·7	144·6	147·4	150·3	153·1	155·9	158·8	161·6	164·4	167·3
6	170·1	172·9	175·8	178·6	181·4	184·3	187·1	189·9	192·8	195·6
7	198·4	201·3	204·1	207·0	209·8	212·6	215·5	218·3	221·1	224·0
8	226·8	229·6	232·5	253·3	238·1	241·0	243·8	246·6	249·5	253·2
9	255·1	258·0	260·8	263·7	266·5	269·3	272·2	275·0	277·8	280·7
10	283·5	—	—	—	—	—	—	—	—	—

TABLE IX. GRAMMES TO OUNCES

(to 2 decimal places)

↓ grammes →	0	1	2	3	4	5	6	7	8	9
0	—	0·035	0·07	0·105	0·14	0·118	0·121	0·125	0·128	0·132
10	0·35	0·39	0·42	0·46	0·49	0·53	0·56	0·60	0·63	0·67
20	0·71	0·74	0·78	0·81	0·85	0·88	0·92	0·95	0·99	1·02
30	1·06	1·09	1·13	1·16	1·20	1·23	1·27	1·31	1·34	1·38
40	1·41	1·45	1·48	1·52	1·55	1·59	1·62	1·66	1·69	1·73
50	1·76	1·80	1·83	1·87	1·90	1·94	1·98	2·01	2·05	2·08
60	2·12	2·15	2·19	2·22	2·26	2·29	2·33	2·36	2·40	2·43
70	2·47	2·50	2·54	2·58	2·61	2·65	2·68	2·72	2·75	2·79
80	2·82	2·86	2·89	2·93	2·96	3·00	3·03	3·07	3·10	3·14
90	3·17	3·21	3·25	3·78	3·32	3·35	3·39	3·42	3·46	3·49
100	3·53	—	—	—	—	—	—	—	—	—

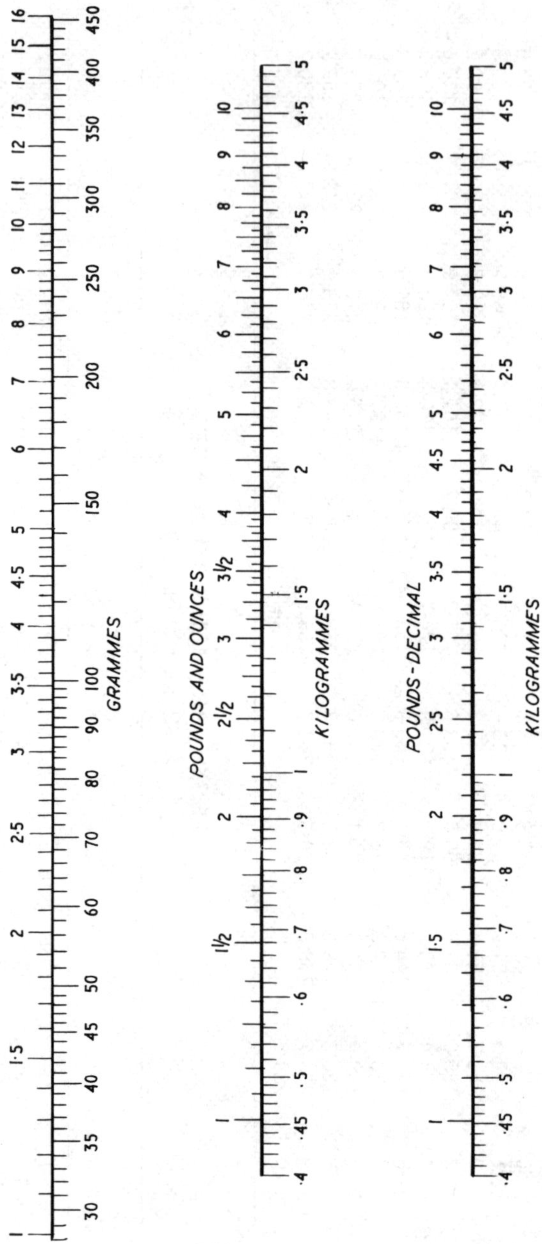
OUNCES
GRAMMES
POUNDS AND OUNCES
KILOGRAMMES
POUNDS - DECIMAL
KILOGRAMMES

TABLE X.

Pounds	Kilogrammes
1	0·454
2	0·907
3	1·361
4	1·814
5	2·268
6	2·722
7	3·175
8	3·629
9	4·082
10	4·536

Kilogrammes	Pounds
1	2·205
2	4·409
3	6·614
4	8·819
5	11·023
6	13·228
7	15·432
8	17·637
9	19·842
10	22·046

6. Density

Density is mass per unit volume, the new ISO standard being kilogrammes per cubic metre (kg/m^3). This does yield particularly unfamiliar values for people used to working in the standard English unit of pounds per cubic foot.

1 kg/m^3 = 0·06243 lb/cu. ft
1 lb/cu. ft = 16·018 kg/m^3

Conversion is, however, quite straightforward. For ordinary purposes it is sufficient to remember that ISO standard density, in kg/m^3, is 16 times density in lb/cu. ft. For example, the metric equivalent density of 8 lb/cu. ft balsa would be 8 × 16 = 128 kg/m^3. Similarly, if the density is quoted in units of kg/m^3, divide by 16 to bring the density figure to lb/cu. ft.

Despite the ISO standard, metric densities may also be quoted in (metric) tonnes per cubic metre, where

1 tonne/m^3 = 62·428 lb/cu. ft
1 lb/cu. ft = 0·016 tonnes/m^3

This is quite a convenient practical unit for tonnes per cubic metre are equal to kilogrammes per cubic decimetre, and also to grams per cubic centimetre.

Liquid densities are commonly quoted in the latter unit (grammes per cubic centimetre); or more correctly as grammes per millilitre (see CAPACITY for the difference).

1 gramme per millilitre = 10·0221 (say 10) pounds per gallon
1 pound per gallon = 0·099779 (say one-tenth) grammes per millilitre

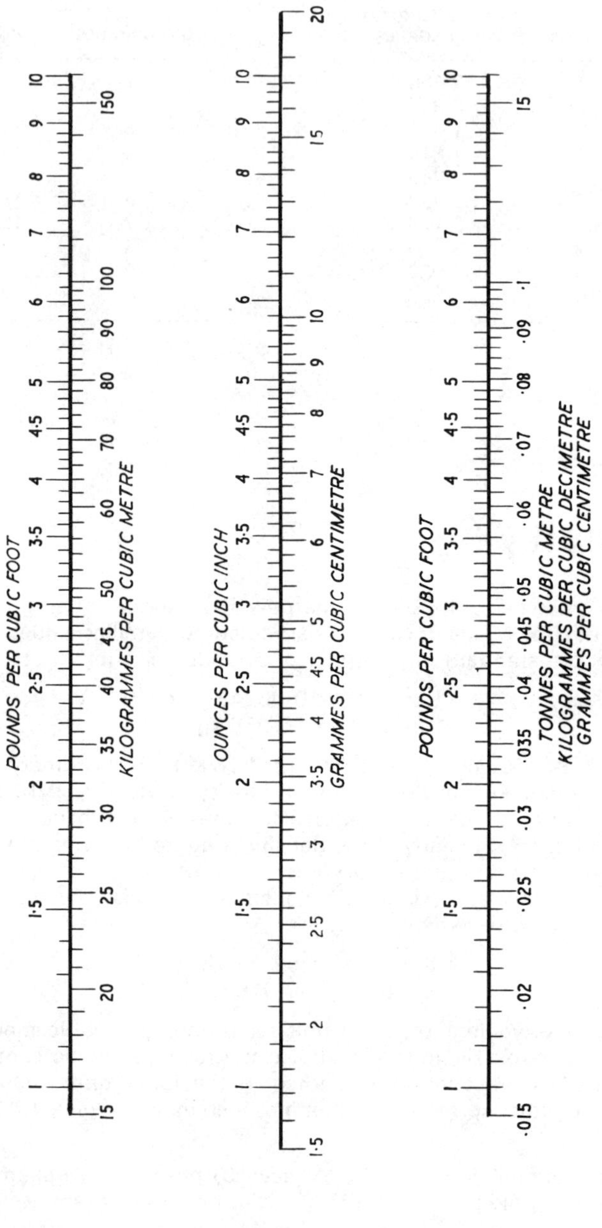
POUNDS PER CUBIC FOOT
KILOGRAMMES PER CUBIC METRE
OUNCES PER CUBIC INCH
GRAMMES PER CUBIC CENTIMETRE
POUNDS PER CUBIC FOOT
TONNES PER CUBIC METRE
KILOGRAMMES PER CUBIC DECIMETRE
GRAMMES PER CUBIC CENTIMETRE

Some confusion can arise when working from Continental specifications since the French word *densité* does not mean the same as English density, but rather *specific gravity*. The French equivalent for density is *mass volumique*.

The use of the term *specific gravity* is also now frowned upon. The international recommendation is that it should be replaced by *relative density*, which has the same numerical value. The corresponding French is *densité relative*, which avoids any confusion with 'density'. Somehow, however, it is felt that the term specific gravity (and its convenient abbreviation SG) will continue to be used for a long time in this country, where it has been standard for a long time (e.g. battery electrolyte condition as determined by hydrometer measurement).

7. Force and Moments (Torque)

It is far easier to appreciate what a 'force' is, and to use its numerical value(s) for calculation, than it is to define the derivation of a unit of force. The latter is defined from the unit of mass by considering the force necessary to give unit acceleration to unit mass. For practical working, all this really means is that the unit of force has the same name as that for mass (or, popularly, 'weight')—see WEIGHTS. Thus a force is said to have a value of so-many pounds (or ounces, or kilogrammes, or grammes). The same conversions as for 'WEIGHTS' then apply.

Academically it is correct to distinguish between a mass ('weight') unit and a force unit.

Thus lb implies pounds *mass* ('weight')

lbf implies pounds *force*
kg implies kilogrammes *mass* ('weight')
kgf implies kilogrammes *force*.

The pound force and the kilogramme force are the old English and metric units, respectively. The new ISO metric unit for force is the Newton (N), defined as kilogrammes × metres per second per second.

1 pound force (1 lbf) = 4·44822 N (5 oz 4·45 N)
1 ounce force (1 ozf) = 0·278 N (5 oz 0·28)
1 Newton (1 N) =0·225 lbf
= 3·6 ozf

Probably the best advice to British engineers is that having got these figures for a reference they can forget about Newtons entirely for ordinary working! In any case the metric *standard* adopted is not the Newton or its sub-multiples, but the *megadyne*. We have then

1 kilogramme (force (1 kgf) = 0·981 megadynes
1 pound force (1 lbf) = 0·456 kfg
= 0·445 megadynes.

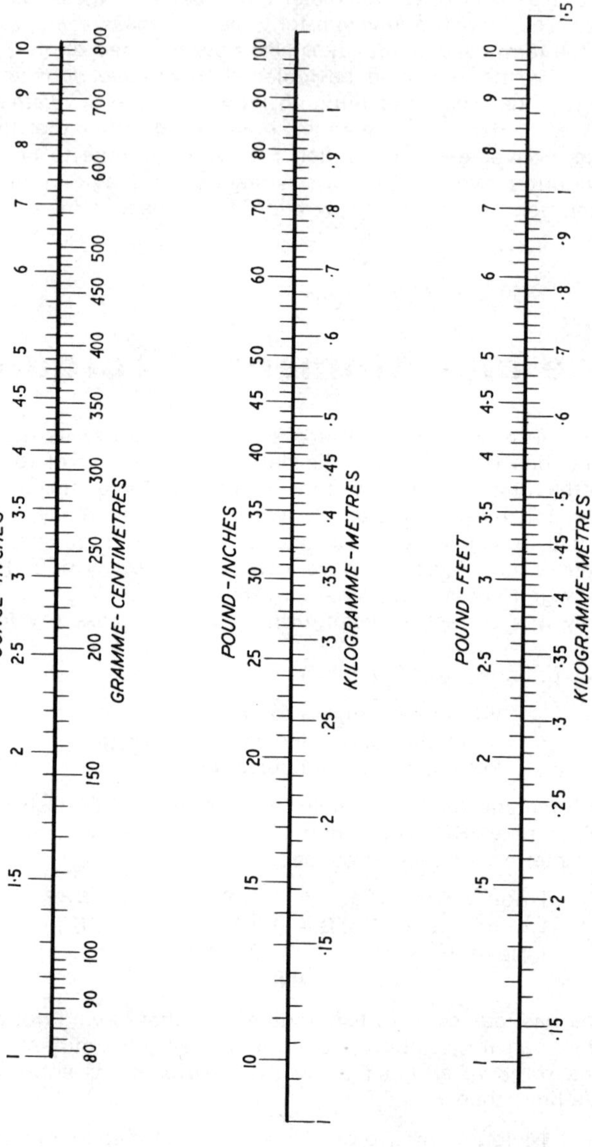
OUNCE-INCHES
GRAMME-CENTIMETRES
POUND-INCHES
KILOGRAMME-METRES
POUND-FEET
KILOGRAMME-METRES

Moral: stick to 'pounds' (lbf) or 'kilogrammes' (kgf) for force units!! And if Newtons or megadynes do crop up, convert back into these more familiar units!

Moments

Moments, of course, are the product of a force and a distance. Familiarly, units are pound-feet (lb-ft, or more correctly lbf-ft); or kilogramme-metres (kg-m, or more correctly kgf-m), or sub-multiples, e.g. gramme-centimetres.

Conversions are fairly straightforward:

1 pound-foot = 0·138255 kilogramme-metres

It follows that

a pound-inch would be 1/12th of this value; and
an ounce-foot would be 1/16th of this value
an ounce-inch would be 1/12 × 1/16 or
1/192 of this value (approximately 1/200th).

1 kilogramme-metre = 7·233 pound-feet
= 86·8 pound-inches
= 1268·7 ounce-inches

Sub-multiples of the metric unit can easily be dealt with by shifting the decimal point, if necessary thinking in two stages. Thus to convert gramme-centimetres into inch-ounces, think first of a gramme being 1/1000th of a kilogramme.

1 gramme-metre = 1/1000th of 1268·7
= 1·2687 ounce-inches.

A centimetre is 1/100th of a metre. Thus

1 gramme-centimetre = 1/100th of a gramme-metre
= 1/100 of 1/2687
= 0·012687 ounce-inches (say 0·0129)

Incidentally, all the units quoted have been written with the 'force' first, then the 'distance'. This is recommended practice, and followed with all metric units. Common usage, however, favours reversing the order with English units. Thus one speaks of moments in terms of foot-pounds, or inch-ounces, which seems to sound better to the engineer. There are probably academic reasons why this is wrong (see WORK AND POWER), but in fact since we are simply multiplying a 'force' by a 'distance' to find a numerical value of a 'moment', it really does not matter which way round the units are quoted. Another 'difference of opinion' is whether a hyphen is used between the two basic units or not, e.g. lbf-ft or just lbf ft, etc. Again it does not seem to matter, but the hyphen is now usually omitted, particularly when using abbreviations.

The ISO standard for moment uses the Newton, of course (since this is standard force unit in that system)—allied to the ISO standard of length, the metre. In other words, under this system the unit for moment is Newton metres (Nm)

1 lbf ft = 1·35582 Nm
1 Nm = 0·7374 lbf ft

Torque

Torque or 'turning effort' is a moment again, so the same units apply. The conversions most commonly required are

1 lb-ft = 0·138 kg-metre
1 kg-metre = 7·233 lb-ft
1 ounce-inch = 77·52 gramme-centimetres
1 gramme-centimetre = 0·0129 ounce-inches

The figures are the same whether words or abbreviations are used! This is largely a matter of convenience. To be clear, engineers often prefer spelling out the units they are using rather than using abbreviations.

8. Work and Power

Work is the product of a force applied and the distance through which it is moved. Thus we have the same basic units as moments. Common usage always puts the 'distance' before the 'force' in English units, e.g. foot-pounds, inch-pounds or inch-ounces. Metric units are quoted the other way round—kilogramme-metres, gramme-centimetres.

The ISO standard groups work, energy and heat together, using the single standard unit the Joule (J). To say the least this sounds confusing, if not misleading at times. However, the Joule is only a Newton metre (Nm), so

1 Joule = 0·737 foot-pound (ft-lbf)
1 foot-pound (ft-lbf) = 1·35582 Joules (J)
1 inch-ounce (in-ozf) = 0·00706 Joules (J)

In more familiar units, remember again that

1 foot-pound = 0·138 kilogramme-metres
1 inch-ounce = 77·52 gramme-centimetres
1 kilogramme-metre = 7·233 foot-pounds
1 gramme-centimetre = 0·0129 inch-ounces

Power is a *rate* of doing work—e.g. torque multiplied by rpm—with the long established unit of the *horsepower.* There is, however, a small difference between the English horsepower and metric horsepower (known also as *cheval vapeur*)

1 horsepower = 1·01387 metric horsepower
1 metric horsepower = 0·98632 horsepower.

This difference is usually small enough to be ignored.

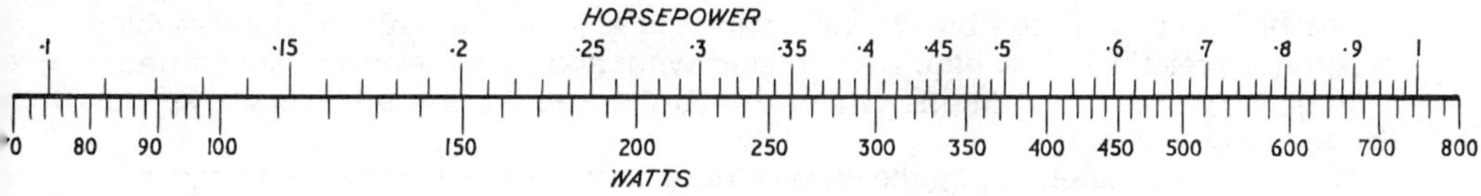

The horsepower is the standard unit for rating the power output of internal combustion engines and other prime movers. It does, however, have different forms of rating, depending on the manner in which horsepower measurement is derived on test. The horsepower *unit* is the same in all cases, the difference between different test ratings being largely a matter of how many ancillaries the engine is driving on test. This affects larger engines only and seldom enters into model practice, except that the actual apparatus used to measure horsepower output may itself imply a 'rating' by virtue of its own unmeasured power absorption.

To provide a direct comparison between i/c engines and electric motors, the latter may also have their power output determined in terms of horsepower. For model sizes, the unit millihorsepower has been used from time to time, but never gained universal acclaim. Rather the preference has been to rate electric motor power in terms of watts, the relationship between watts and horsepower being

1 horsepower = 746 watts = 0·746 kilowatts
1 kilowatt = 1·34 horsepower
1 watt = 0·00134 horsepower = 1·34 millihorsepower

Fractional horsepower values (i.e. under 1 horsepower) should be converted in terms of equivalent watts. For horsepower values of 1 or greater, the equivalent should be rendered in terms of kilowatts.

e.g. 0·8 HP = 600 watts
but 8 HP = 6 kilowatts

Note here that the power rating refers to the *output* of the motor in terms of electrical power, or watts. It is inevitably lower than the *input* power (volts times amps).

Power output (in watts) = input watts × overall efficiency.

Efficiency can range from as high as 90 per cent or better in the case of very large motors, down to less than 10 per cent on very small eiectric motors operating under unfavourable conditions.

The practice of determining electric motor power in terms of watts can, therefore, be misleading at times. What is referred to as a '20 watt motor', for example, may be designed for a 20 watt *input*. Its actual power will depend entirely on its overall efficiency. Thus if its efficiency is 50 per cent, then its *power* rating will be 10 watts.

The difference can be considerable, particularly in model sizes of motors which may have a relatively low efficiency to start with; and they may be operated under a range of working conditions which can further modify the efficiency and thus the actual power output.

Nevertheless, determining the power output of an electric motor in terms of watts is fairly logical, provided it is explicit that this is the power *output* and not the power *input*. Watts output can readily be turned into horsepower equivalent by straightforward conversion.

The ISO system goes one stage further in adopting the *watt* as the standard unit for *all* power measurement, including i/c engines and other prime movers as well as electric motors. Thus a 1 horsepower glow motor would have an ISO power of 746 watts; and a 100 horsepower car engine an ISO power of 74 600 watts or 74·6 kilowatts.

The immediate reaction of nearly all mechanical engineers is instant rejection of such a single standard. It does not make sense for reciprocating engines, etc.—except to the academic type where the watt is a logical power unit since it is equal to a Joule per second. And a Joule is a Newton metre. Take it from there!

Meantime, the horsepower as a unit for mechanical power (as opposed to electric motor power) will undoubtedly continue to be used for a long time. And conversions, when necessary, are simple enough. Use the close approximation:

$$1 \text{ horsepower} = \tfrac{3}{4} \text{ of a kilowatt}$$
$$1 \text{ kilowatt} = 1 \text{ and } 1/3\text{rd horsepower}$$

9. Pressure and Stress

A *pressure* unit is defined as force per unit area. *Stress* is defined in a similar way, so the two units are the same. The well-accepted English unit is pounds per square inch, commonly abbreviated to *psi*, but also sometimes written out—e.g. lb/sq. in. Strictly speaking, it should now be written lbf/in^2—but 'psi' is well enough understood to continue to be used in defiance of overall standardisation!

The corresponding metric unit is kilogrammes per sq. cm (kg/per sq. cm, or more correctly kgf/cm^2, although kg/cm^2 is good enough).

$$1 \text{ psi } (1 \text{ lb/in}^2) = 0{\cdot}07 \text{ kg/cm}^2$$

That is easy enough to remember, and becomes quite familiar if you work a lot with pressure or stress figure. 100 psi = 7 kg/cm^2; 1000 psi = 70 kg/cm^2; and so on.

$$1 \text{ kg/cm}^2 = 14{\cdot}2233 \text{ psi } (\text{lb/in}^2)$$

This is not so easy to remember, but there is a clue. 1 kg/cm^2 is 'atmospheric pressure', or roughly 14·4 lb/in^2 (see also later remarks on 'atmospheric pressure').

Higher stress values (and occasionally pressures, but not to the same extent), are often given in *tons per square inch* in English units. This, in fact, is not a very easy unit to use, for it normally has to be multiplied by 2240 for working out answers anyway, i.e. to get back to *pounds* force. And it is only applied to 'long' tons, not 'short' (2000 pounds) tons.

The metric equivalent in this case is kilogrammes per *square millimetre* (not square centimetres, which would lead to unwieldy numerical values).

1 ton per sq. in = 0·16 kg/sq. mm (approximate—but close enough for most working).
1 kilogramme per sq. mm = 0·635 tons per sq. in.

Kilogrammes per sq. mm multiplied by 100, of course, gives kg/cm^2.

All these reasonably rational units go by the board when the ISO standard unit is considered for pressure, which is Newtons per square metre (N/m^2). Apart from the fact that this brings in the disliked 'Newton', the resulting unit yields such low quantitative values at practical pressure levels that it is quite ridiculous. To overcome this the basic SI unit has to be multiplied by 1000 (i.e. kiloNewtons/square metre or kN/m^2); or 1 000 000 (i.e. megaNewtons/square metre, or MN/m^2).

Even this does not work out properly and an actual unit factor of 100 000 is better, i.e. 10^5 N/m^2. This is nominally the same as atmospheric pressure, or the unit known as the *bar*. The older metric unit of kg/cm^2 is also about the same as atmospheric pressure, as noted above, so the bar and the kg/cm^2 are basically the same in value—but not quite!

Standard atmospheric pressure = 14·7 lb/sq. in.
a bar = 14·5 lb/sq. in. (approx.)
1 kg/cm^2 = 14·4 lb/sq. in. (approx.)

All the above are close approximations, not exact. Even so, the bar, lb/sq. in. and kg/cm^2 all have slightly different values, and differ from 'standard' atmospheric pressure.

But that is not the end. There is also another 'standard' value for atmospheric pressure, known as the *technical atmosphere*, equivalent to 14·2 lb/sq. in. (approx); and still *another* pressure unit based on this known as the kilopond per square centimetre (kp/cm^2).

1 kp/cm^2 = 14·2 lb/sq. in.

The 'rational approach seems fairly straightforward. If we are used to working in units of lb/sq. in, it makes sense to continue to use this unit—and call it 'psi' if preferred. *Any* of the metric units based on 'atmospheric pressure', whether designated *bar*, atmospheres (Am), technical atmosphere (At) or kiloponds per square centimetre (kp/cm^2), can be considered *the same as kilogrammes per sq. cm* (kg/cm^2), when the easy conversion applies:

1 psi = 0·07 kg/cm^2
1 kg/cm^2 = 14·22 psi

This will be satisfactory for most 'engineering' calculations, but can fall down where very small differences in pressure are involved. In this case the slide rule

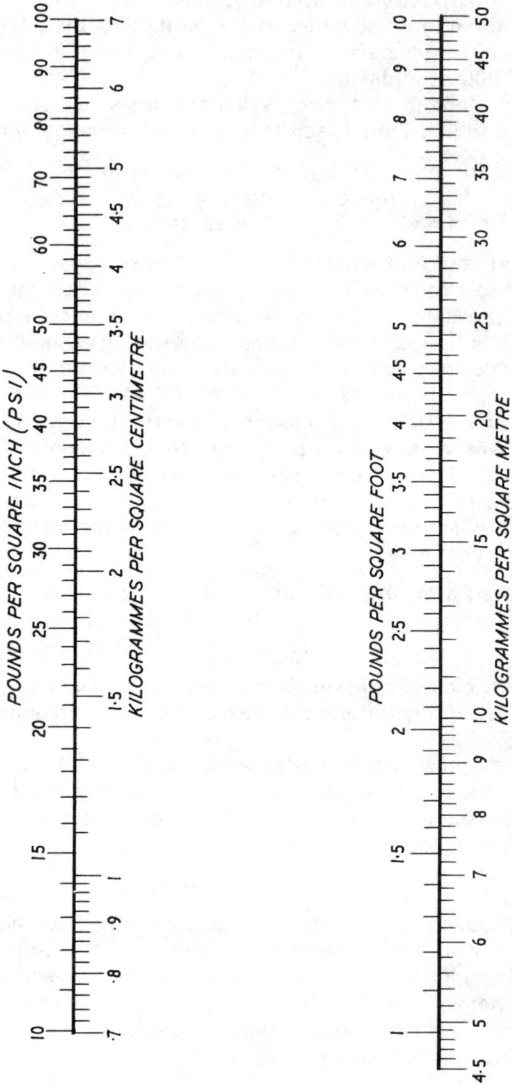
POUNDS PER SQUARE INCH (P S I)
KILOGRAMMES PER SQUARE CENTIMETRE
POUNDS PER SQUARE FOOT
KILOGRAMMES PER SQUARE METRE

will have to come out to convert on the basis

$$1 \text{ psi} = 68{\cdot}9476 \text{ millibars}$$
$$1 \text{ bar} = 14{\cdot}2 \text{ psi}$$

For nearly all ordinary workings, the instant conversion scales will provide all the solutions necessary.

10. Miscellaneous Units

For 'metrication troubles' with various other units, see if the quantity concerned appears under the following sub-headings and follow the general guidance given. If not, refer to the summary of SI units—see Appendix 2.

Temperature

The Fahrenheit and Centigrade temperature scales have existed alongside each other for many years. Under the proposed International System of Units both disappear, in favour of temperature in degrees *Celsius*. Not to worry. This—i.e. degrees Celsius or °C—*is* the original Centigrade scale. Only the name has been changed—mainly because 'grade' in French technical language means 1/100th of a right angle, and so 'centigrade' means 1/10 000th of a right angle.

$$°C = 5/9\text{ths of } (°F - 32)$$
$$°F = 9/5 \textbf{ times } °C \textbf{ plus } 32$$

Temperatures are also sometimes quoted as *absolute* values, meaning that they are measured from absolute zero of temperature. The *absolute Fahrenheit scale* is known as the *Rankine* scale (°R) where

$$°R = °F + 460$$

The corresponding *absolute Centigrade scale* is degrees *Kelvin* (°K)

$$°K = °C + 272$$

Speed (Velocity)

Time is a common unit in all systems—the second being the basic unit, with the minute and hour as a convenient and still universally accepted multiples. Speed (or velocity) per minute is less used, unless it gives a more convenient numerical figure (i.e. not too small, nor too large).

The following should provide all the conversion factors necessary.

Miles per hour	*Feet per minute*	*Feet per second*	*Centimetres per second*	*Kilometres per hour*
1·0	88	1·4667 (say 1·5)	44·704	1·609
0·0114	1·0	0·0167	0·508	0·0183
0·682	60·0	1·0	30·48	1·097
0·022	1·968 (say 2·0)	0·033	1·0	0·036
0·621	54·68	0·911	27·78	1·0

The UK nautical mile of 6080 feet, incidentally, is approximately 0·06 per cent greater than the international nautical mile (standardised as 1852 metres, exactly).

1 UK knot = 1·15 mph (approx)
= 1·853 kilometres/hour

Acceleration

Basic 'equivalents' are feet per second per second (ft/sec^2) and metres per second per second (m/sec^2), where

$$1 \text{ ft/sec}^2 = 0{\cdot}3048 \text{ m/sec}^2$$
$$1 \text{ m/sec}^2 = 3{\cdot}28 \text{ ft/sec}^2$$

There is the possibility, though, that you may find acceleration quoted in Gal, or decimal multiples of a Gal in some metric workings. This is not a SI unit, but it does have some use in Continental work. 1 Gal is simply 1 cm/sec^2

Angular Measure

The standard unit for an angle is the *radian,* where 2π radians equal 360 degrees. This is a universal mathematical function, but for normal use degrees are more convenient than radians (or milliradians, as the SI system advocates suggest for smaller units).

If you ever need to convert,

radians times 57·3 = degrees
degrees times 0·0175 = radians

Minutes and seconds remain acceptable sub-divisions of a degree, but you might also come across the unit 'grade'. This is a sub-division equal to $\pi/200$ radians (see also notes on 'Temperature').

Viscosity

Modellers are not generally concerned with liquid viscosities. In any case, most of the practical viscosity scales are arbitrary ones, such as Redwood No. 1 (British), Saybolt Universal Seconds or S.U.S. (American) and Engler degrees (Continental Europe). 'SAE viscosities' are also quoted for lubricating oils, but these are not viscosity values at all, merely a nominal specification giving viscosity *limits* at particular temperatures.

Standard viscosity units adopted from the cgs system are the *centipoise* for dynamic viscosity; and *centistoke* for kinematic viscosity.

1 centipoise = Newton seconds per metre squared* $\times 10^{-3}$
1 centistoke = metres^2 per second* $\times 10^{-6}$

If you ever do need to work out equivalent 'English' engineering units:

Dynamic viscosity:

1 Newton second/metre^2 = 0·021 pound sec/ft^2 (approx)
1 centipoise = 2·08894 $\times 10^{-5}$ pound sec/ft^2

Kinematic viscosity:

1 metre^2/sec = 10·764 ft^2/sec
1 centistoke = 5·58 in^2/hour

Frequency

The general tendency is now to adopt the *Hertz* (Hz) as the standard unit for frequency rather than *cycles per second,* the two meaning exactly the same thing (i.e. 1 hertz = 1 cycle per second or 1 cps). Decimal multiples are used for larger values, e.g.

1 kilohertz (1 kHz) = 1000 Hz
1 megahertz (1 MHz) = 1 000 000 Hz
1 gigahertz (1 GHz) = 1 000 000 000 Hz

Electrical Units

Electrical units are not greatly affected by metrication since these are already based on the metric system, and existing SI and CGS units. The ampere (current), volt (emf or potential), coulomb (electric charge), ohm (resistance), farad (capacitance) and henry (inductance) remain standard. About the only two important changes to note are:

Magnetic Field Strength, which has hitherto usually been quoted in *oersteds,* is now defined in units of *amperes per metre*. The equivalent value is given by multiplying oersteds by $1000/4\pi$. The new unit has no specific name.
Magnetic Flux Density, which has hitherto been quoted in *gauss* is now defined in terms of the new unit *tesla* (T). To convert gauss to tesla, divide by 10 000.

Appendix 1. Prefixes—Multiples and Sub-multiples

Prefix	*Symbol*	*Factor by which the unit is multiplied*		*Example*
tera	T	one million million (billion†)	$= 10^{12}$	
giga	G	one thousand million	$= 10^{9}$	
mega	M	one million	$= 10^{6}$	megawatt (MW)
kilo	k	1000	$= 10^{3}$	kilometre (km)
hecto	h	100	$= 10^{2}$	
deca	da	10	$= 10^{1}$	decagramme (dag)
deci	d	0·1	$= 10^{-1}$	decimetre (dm)

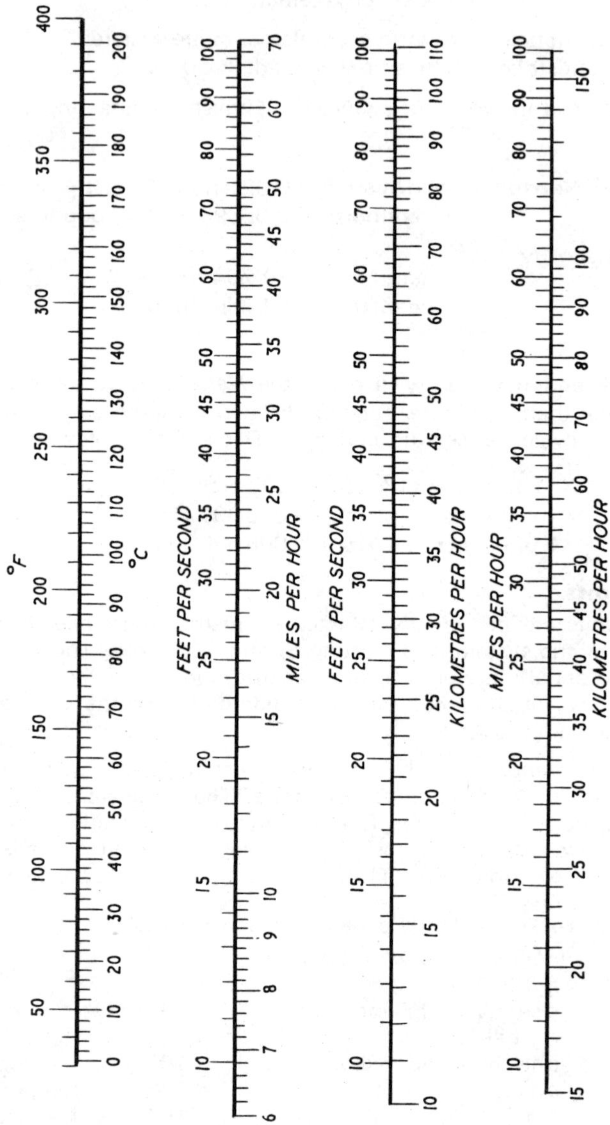
°F
°C
FEET PER SECOND
MILES PER HOUR
FEET PER SECOND
KILOMETRES PER HOUR
MILES PER HOUR
KILOMETRES PER HOUR

Prefix	*Symbol*	*Factor by which the unit is multiplied*		*Example*
centi	c	0·01	$= 10^{-2}$	centimetre (cm)
milli	m	0·001	$= 10^{-3}$	milligramme (mg)
micro	μ	0·000 001 (one millionth)	$= 10^{-6}$	microsecond (μs)
nano	n	one thousand-millionth	$= 10^{-9}$	nanometre (nm)
pico	p	one million-millionth (one billionth†)	$= 10^{-12}$	picofarad (pF)
femto	f	one thousand-million-millionth (one thousand-billionth†)	$= 10^{-15}$	
atto	a	one million-million-millionth (one trillionth†)	$= 10^{-18}$	

† English usage; American usage differs

Note: the above are specific to any decimal or metric system, but the same prefixed may be used with English units, e.g.

millihorsepower = 1/1000th of a horsepower
megaton = a million tons

Appendix 2. The International (SI) System of Units

Quantity	*Unit*	*Symbol*
length	metre	m
mass	kilogramme	kg
time	second	s
electric current	ampere	A
temperature	degree Kelvin	°K
luminous intensity	candela	cd

Supplementary and derived units:

Quantity	*Unit*	*Symbol or abbreviation*
Plane angle	radian	rad
Solid angle	steradian	sr
Area	square metre	m^2
Volume	cubic metre	m^3
Angular velocity	radian per second	rad/s
Velocity	metre per second	m/s
Acceleration	metre per second squared	m/s^2
Angular acceleration	radian per second squared	rad/s^2
Frequency	hertz	Hz (s^{-1})
Rotational frequency	reciprocal second	s^{-1}

Quantity	*Unit*	*Symbol abbreviation*
Density	kilogramme per cubic metre	kg/m^3
Momentum	kilogramme metre per second	kg m/s
Angular momentum	kilogramme metre squared per second	$kg\ m^2/s$
Moment of inertia	kilogramme metre squared	$kg\ m^2$
Force	Newton	N $(kg\ m/s^2)$
Moment of force	Newton metre	N m
Pressure, stress	Newton per square metre	N/m^2
Viscosity: kinematic	metre squared per second	m^2/s
dynamic	Newton second per metre squared	$N\ s/m^2$
Work, energy, quantity of heat	joule	J (N m)
Power, heat flow rate	watt	W (J/s)
Temperature (customary unit)	degree Celsius	°C
Temperature interval	degree Kelvin, degree Celsius	°K, degC, °C
Thermal coefficient of linear expansion	reciprocal degree (Kelvin or Celsius)	$degC^{-1}$, $°K^{-1}$, $°C^{-1}$
Density of heat flow rate	watt per square metre	W/m^2
Thermal conductivity	watt per metre degree	W/m degC
Coefficient of heat transfer	watt per square metre degree	W/m^2 degC
Heat capacity	joule per degree	J/degC
Specific heat capacity	joule per kilogramme degree	J/kg degC
Entropy	joule per degree Kelvin	J/°K
Specific entropy	joule per kilogramme degree Kelvin	J/kg °K
Specific energy, specific latent heat	joule per kilogramme	J/kg
Quantity of electricity, electric charge	coulomb	C (A s)
Electric tension, potential difference, electromotive force	volt	V (W/A)
Electric field strength	volt per metre	V/m
Electric resistance	ohm	Ω (V/A)
Electric capacitance	farad	F (A s/V)
Magnetic flux	weber	Wb (V s)
Inductance	henry	H (V s/A)
Magnetic flux density	tesla	T (Wb/m^2)
Magnetic field strength	ampere per metre	A/m
Magnetomotive force	ampere	A
Luminous flux	lumen	lm (cd sr)
Luminance	candela per square metre	cd/m^2
Illumination	lux	lx (lm/m^2)